Freddy Ahadi Ndiweno

Mothers' knowledge, attitudes and practices

Freddy Ahadi Ndiweno

Mothers' knowledge, attitudes and practices

for supplementary feeding in the BAGIRA health zone

ScienciaScripts

Imprint

Cover image: www.ingimage.com

This book is a translation from the original published under ISBN 978-620-6-71251-0.

Publisher:
Sciencia Scripts
is a trademark of
Dodo Books Indian Ocean Ltd. and OmniScriptum S.R.L publishing group

120 High Road, East Finchley, London, N2 9ED, United Kingdom
Str. Armeneasca 28/1, office 1, Chisinau MD-2012, Republic of Moldova, Europe
Printed at: see last page
ISBN: 978-620-7-61909-2

MOTHERS' KNOWLEDGE, ATTITUDES AND PRACTICES REGARDING COMPLEMENTARY FEEDING IN THE BAGIRA HEALTH ZONE

BY AHADI NDIWENO FREDDY

TABLE OF CONTENTS

EPIGRAPH 3

DEDICATION 4

THANKS.. 5

SUMMARY 6

INTRODUCTION 7

OBJECTIVES...13

INTEREST OF THE WORK...14

CHAPTER I 15

CHAPTER II 22

CHAPTER III...27

CONCLUSION 39

RECOMMENDATIONS 41

BIBLIOGRAPHICAL REFERENCES 42

APPENDIX 44

EPIGRAPH

We make science out of facts, just as we make a house out of stones, but an accumulation of facts is not a science any more than a pile of stones is a house. Pyrite is to the chemical industry what bread is to human nutrition.

"Paul Truchot

DEDICATION

To my very dear parents NDIWENO NTUMULO Louis and SIFA MULEMANGABO Béatrice.

To my dear and adorable brothers and sisters: MICHEL NDIWENO, MICHELINE NDIWENO, JOHN MULEMANGABO, FIKIRINI NDIWENO Naomi, SOLANGE NDIWENO Gabriella, FURAHA NDIWENO Docile, BILUGE NDIWENO Daniella, JIBU NDIWENO Carine, BARAKA NDIWENO Ornella, MUGOLI NDIWENO, MULUMEODERHWA NDIWENO Franck, ROSINE NDIWENO, RUTH NDIWENO, JOLIE NDIWENO, THERESE NDIWENO.

AHADI NDIWENO Freddy

THANKS

We would like to take this opportunity to express our gratitude to all those who contributed **to** this study, for their commitment and availability at every stage.
To the academic and administrative authorities of UOB for having contributed to our scientific training. To all the staff of the Faculty of Pharmaceutical Sciences and Public Health of the official University of Bukavu in general for having provided, without interruption, our training in Public Health during these three years that we have just spent at the official University of Bukavu and in particular to Mr ASIMA KATUMBI Florentin, supervisor of the work, who in spite of multiple occupations agreed to direct us.
Our warmest thanks and deepest gratitude also go to the NDIWENO family, for their love and all-round education, which I never cease to benefit from.
We would also like to express our heartfelt gratitude to our closest friends, in particular MUHANZI ZIHALIRWA Lucien, NAGAZIRE IRENGE Victoire, MUPENDA MWISHA Anicet, USHINDI PASCASIE, ZEZI NGWASI Prisca.
To our friends and comrades whose love and sympathy have always marked us: GERMAIN MURHULA, KASANGA AMISI, CHIHEBEY BABA-NEEMA Cossta, MUGISHO SAFARI David, SHUKURU BISIMWA Judith, ANSIMA MATABARO, SCOLASTIQUE MWATI, AJUAMUNGU MUSHOSI Siméon.
We would like to express our gratitude to all those who contributed in one way or another to this work.

SUMMARY

Introduction: Feeding children over the age of 6 months remains a real public health problem. The aim of this study is to assess the level of knowledge, attitudes and practices of mothers in the Bagira health zone with regard to complementary feeding.

Methodology: This was a descriptive cross-sectional study. The LUNCH formula enabled us to find a sample of 384 breastfeeding mothers. Data were collected using a survey questionnaire in the Bagira health zone.

Results: Our study showed that 99% of respondents had already heard of complementary feeding. The main source of information was the radio (60.03%). Almost all of the respondents knew that complementary feeding is the practice of introducing complementary foods to breast milk at 6 months. Ordinary porridge was identified as the very first food to be introduced by 9 out of 10 respondents, i.e. 91.8%, and almost all (97.9%) mentioned the age of over 24 months as the weaning age.

Conclusion: It is important to improve women's knowledge of the composition of food to be given to children so that they have a balanced diet.

Key words: Knowledge, attitudes, practices, supplementary feeding, Bagira health zone.

INTRODUCTION

I.1. BACKGROUND

Countless efforts are being made on an international scale to ensure sufficient, healthy food for all mankind. However, it has to be admitted that this challenge has not yet been met. Malnutrition is defined as a pathological state resulting from a relative or total deficiency or excess of one or more essential nutrients, whether manifested clinically or detectable only by biochemical, anthropometric or physiological analyses **(WHO, 2019)**. While it is true that malnutrition due to excess is a growing problem, we should note that malnutrition due to deficiency is the most widespread. More than a billion people worldwide suffer from malnutrition (one in 10), most of them in developing countries. Around two billion people are deficient in one or more micronutrients. The main victims are women of childbearing age and, above all, children, especially those under the age of five **(PRONANUT, 2010).**

The **WHO** recommends that newborns should be put to the breast immediately after birth (30 minutes), that they should be exclusively breastfed for up to 6 months and that they should be diversified with safe and appropriate foods from 6 months onwards **(WHO, 2019)**. Efforts still need to be made in our country to improve feeding practices among young children in order to guarantee them better nutritional status, reduce morbidity and mortality rates linked to inappropriate feeding practices and thus ensure that these children have a more secure future at school and at work. This is why we decided to undertake this study of mothers' knowledge, attitudes and practices with regard to early childhood nutrition in the Bagira health zone.

I.2. ISSUES

The most acute form of malnutrition is "wasting", which carries the most immediate risk of mortality. Worldwide, 51 million children under the age of five suffer from acute malnutrition, 19 million of whom are affected by the most severe form. Despite significant efforts, only 3 million of them have access to treatment **(INFO-ECHO, 2018) .**

For the third year running, hunger in the world has gained ground. The absolute number of undernourished people, i.e. those suffering from chronic food deficiency, rose to almost 821 million in 2017, compared with around 804 million in 2016. This is almost ten years ago. The proportion of undernourished people in the world population - the Prevalence of Undernourishment (PoU) - could have reached 10.9% in 2017 **(HIMOURA, 2020).**

Malnutrition is responsible for the death of more than **6 million** children under the age of five worldwide. In young children, it is responsible for serious physical deficiencies as well as major cognitive problems that will hinder their educational and professional future to a greater or lesser extent **(INFO-ECHO, 2018).**

Worldwide, 149.2 million children under the age of 5 are stunted, 45.4 million are wasted and 38.9 million are overweight. More than 40% of all men and women (2.2 billion people) are now overweight or obese. Some countries are making encouraging progress.

The 2017 FAO report noted that after declining for several years, the prevalence of hunger had risen again in Africa. The most recent data presented in this year's Regional Overview confirms that this trend is continuing, with Central and West Africa most affected. Today, one-fifth of Africans are undernourished, representing 257 million people **(HIMOURA, 2020).**

642 million in Asia and the Pacific, **265 million** in sub-Saharan Africa**, 53 million in Africa and the Middle East.**

in Latin America and the Caribbean, and **42 million** in North Africa and the Middle East.

In Mali, according to the 2006 Demographic and Health Survey, 37% of children under the age of 5 are chronically malnourished, 15% are acutely malnourished and 27% are underweight (**WHO, 2006**).

The Democratic Republic of Congo **(DRC)** is one of ten countries that account for 60% of the global burden of wasting in children under 5. The country has high rates of acute malnutrition: 6.5 per cent global acute malnutrition (GAM) and 2 per cent severe acute malnutrition (SAM) according to the latest MICS 2018. Two of the 26 provinces have a SAM prevalence above the critical and emergency threshold of 5% (Ituri and Nord Ubangi). Twelve of the 26 provinces have a SAM prevalence of at least 2%. Taking into account the COVID-19 context, the Nutrition Cluster has estimated that in 2021 there will be 3.8 million children under the age of five affected by acute malnutrition, including 1.1 million suffering from severe acute malnutrition. The prevalence of chronic malnutrition has remained very high and unchanged. According to the latest MICS 2018 survey, 41.8% of children under five suffer from stunting (chronic malnutrition) in the DRC. The country ranks 8th in the world for stunting rates. It is estimated that more than 7 million children under the age of 5 suffer from stunting in 25 of the 26 most affected provinces in the DRC (prevalence > 30%, critical threshold according to the WHO) and 3 provinces with a prevalence of more than 50% (Kwango, Kasai Central and Sankuru). In 2019, the nutritional situation remained worrying in eleven territories, where nutrition and mortality surveys conducted in seventeen territories in 10 provinces (Sud-Kivu, Maniema, Kwango, Lomami, Sankuru, Kasai Oriental, Tanganiyka, Tshuapa, Mai Ndombe, Kwilu) showed levels of global acute malnutrition twice as high as the established threshold (5%). Of the seventeen regional nutritional surveys carried out, eleven revealed a high prevalence of global acute malnutrition (GAM) ranging from 16.5% to 10.1%, and from 2.7% to 5% for the prevalence of severe acute malnutrition (SAM). In 2020, according to the results of SMART surveys carried out in 27 health zones in 9 provinces - Equateur (3), Haut Lomami (1), North Kivu (14), South Kivu (2), Kwango (1), Ituri (1), Mai Ndombe (2), Tanganyika (2), Kasai Oriental (2) - 8 health zones showed levels of global acute malnutrition above the established WHO threshold of 10%: Kamina in Haut Lomami Province (SAM: 6.3%, GAM: 15.7%); Bikoro in Equateur (SAM: 4%, GAM: 17.4%); Mwenga in South Kivu (SAM: 3.4% and GAM: 14.3%); Boko in Kwango (SAM: 3%, GAM: 15.8%); Drodro in Ituri (SAM: 5.9%, GAM: 14.3%); Tandembelo in Mai Ndombe (SAM: 2.4% and GAM: 14%); Yumbi in Mai Ndombe (SAM: 3.6% and GAM: 14.4%); Nzaba in Kasai Oriental (SAM: 6.9% and GAM: 17.9%). It should be noted that, in 2020, few nutritional surveys were carried out in zones under nutritional alert due to a lack of funding. Analysis of the nutritional situation in 2020 according to the 4 quarterly Bulletins No 39, 40, 41 and 42 of the food security and early warning nutritional surveillance system (SNSAP), supported by UNICEF/USAID Food for Peace, shows an increase of more than 100% in the number of alerts issued by the health zones compared with the same periods (Quarter 4) in 2019. 238 alerts were declared for 183 health zones in 2020 compared with 178 during the 4 quarters of 2019, which represents 33% more alerts throughout the country. In addition, 43 health zones had 2 or 3 alerts. It should be noted that the provinces of Equateur (19%), Kasai Oriental (13%), Kasai Central (13%), Kwango (11%), Maniema (11%), Tshuapa (11%) and Kasai (8%) accounted for more than 70% of alerts during the 4 quarters (Q1, Q2, Q3 and Q4) of 2020. UNICEF is the lead agency for nutrition in the DRC and the main partner supporting the government's efforts to strengthen and scale up the programme to manage cases of severe acute malnutrition in the Democratic Republic of Congo_ **(UNHCR,**

2020)... The Democratic Republic of Congo (DRC) has the highest number of people suffering from acute food insecurity in Africa, due to the persistence of violence by armed groups in its eastern part, according to the United Nations, which plans to help 8.8 million people with 1.88 billion dollars in 2022 (UN, 2022) .27 million people in the Democratic Republic of Congo, around a quarter of the country's population, have been facing acute food insecurity since September 2021, according to a note from the UN Office for the Coordination of Humanitarian Affairs (OCHA) **(UN, 2022).**

According to the WHO classification, the results of the two surveys show that global acute malnutrition is considered to be "alert". Moreover, the thresholds for intervention in the DRC, as defined by the national nutrition policy, are GAM> 11% and SAM> 2%. According to the P/T index expressed as a z-score, the prevalence of global acute malnutrition (GAM) is 11.9% (8.9-15.6) in Kabare and 12.7% (10.3-15.5) in Walungu. Severe malnutrition exceeds the emergency thresholds in both areas. It is 2.7% (1.5- 4.8) in Kabare and 2.8% (1.7- 4.8) in Walungu. Overall chronic malnutrition in Kabare and Walungu is 53.0% (46.6-59.3) and 31.6% (27.4-36.1) respectively, and exceeds emergency thresholds (MC >30%) **(PRONANUT, 2022).**

According to the 17th round of the IPC, Kabare and Walungu are among the territories in the province of South Kivu that are affected by acute food insecurity and are classified as being in a crisis phase. The presence of armed groups operating in these areas and limited access to agro-pastoral zones are thought to contribute to this situation. **(PRONANUT, 2022).**

The EFSA survey carried out in May 2019 by the WFP also reveals that 86% of households in South Kivu have poor and limited food consumption. The various sources used during the analysis indicate that households use more than 65% of their income to buy food. This results in high rates of households with poor and limited food consumption who resort to severe survival strategies to cope with their difficulty in accessing food **(PRONANUT , 2022).**

South Kivu province is estimated to have around 104,617 children suffering from severe acute malnutrition (SAM), according to the 2019 humanitarian needs analysis for the province. The two territories of Kabare and Walungu account for around a quarter of SAM cases (24617 cases), or 23.5% of SAM children in the province. Although these two territories are expected to have a large number of cases of severe acute malnutrition in 2019, the geographical coverage of health facilities supported by partners in the management of severe acute malnutrition remains low. Only 4 of the 9 health zones are supported by two partners, MDA and INTERSOS, which are present in both territories, and UNICEF is responsible for supplying nutritional inputs. With regard to the management of cases of moderate acute malnutrition (MAM), only one health zone in Kabare is supported by the WFP **(HIMOURA, 2020).**

The management of acute malnutrition remains a challenge in most health zones in the Kabare and Walungu territories, where women and children are most affected. The results of the latest surveys carried out by PRONANUT in these two territories date back to April 2013. The prevalence of GAM was estimated at 12.4% (8.8-17.1) with a severe form of 2.1% (0.9-4.9) in Kabare, while in Walungu it was 12.4% (102-149) with a severe form of 4.8% (3.5-6.6) **(UNHCR, 2020).**

In view of the above, the following questions will guide our research:

- How much does the population of the BAGIRA health zone know about malnutrition?
- What are the causes and consequences of malnutrition?

OBJECTIVES

General objective

- Determine the population's level of knowledge about malnutrition

Specific objectives

- To assess the level of practice of mothers in feeding their children in the BAGIRA health zone.

- Identify the various causes and consequences of malnutrition in the population of the BAGIRA health zone

INTEREST OF THE WORK

As malnutrition is a public health problem, this study will provide an opportunity to understand the various causes and factors that influence malnutrition in the population. From a scientific point of view, as a member of this population, the study helps to give a signal about the health problems of the population in this environment. The study will also serve as a guide for other researchers interested in this topic.

CHAPTER I

LITERATURE REVIEW

I.1. THEORETICAL REVIEW IN GENERAL

I.1 BACKGROUND

Definition of concepts

• **Knowledge, attitudes and practices (KAP)**: the KAP study is a strategic tool for identifying the educational needs of a specific target group. It assesses three points: the level of comprehensive knowledge, the attitudes motivating the behaviour and the **preventive** and management practices of the target populations **(ESSI et al, 2013)**.

• **Knowledge**

Exact idea of a reality, its situation, meaning, characteristics and functioning (**Le Petit Larousse, 2009).**

• **Attitudes** Behaviour we adopt in given circumstances (**Le Petit Larousse, 2009).**

• **Practices** This is the way we usually act (**Le Petit Larousse, 2009).**

• **Population**: in everyday language, a population is a group of people or inhabitants occupying a given space or territory (country, province, town, health district, health zone, health area, etc.). In statistics, a population is defined as a set of elements with the same characteristics **(Célestin K, 2021).**

• **Health zone:** this is a well-defined geographical **entity** (maximum diameter 150km) contained within the boundaries of an administrative territory/commune, comprising a population of at least 100,000 inhabitants (made up of socioculturally homogeneous communities) with health services at 2 interdependent levels (health centres at level 1[er] and a General Hospital of Reference (HGR) at level 2), and shortened by a health centre **(Ministry of Health, 2006).**

- **Alimentation** Action of providing or taking food (**Le Petit Larousse, 2009).**

- **Supplementary power supply**

Process set in motion when breast milk ul or formula milk alone is no longer sufficient to meet an infant's nutritional needs. As a result, other foods and liquids must be added to the breast milk or breast milk substitute. The age range of infants targeted by complementary feeding is generally 6-23 months **(UNICEF, 2019).**

Malnutrition is the result of a nutritional problem, starting with a mother's feeding habits during the gestation period, right up to the birth of her child. The good habits of eating well and feeding your baby breast milk must be maintained if you want to keep your child in good health, since it is quite possible for the child's health to deteriorate even after birth, something that some mothers pretend to ignore. From the period of breastfeeding (infancy) to the age of gradual introduction of food (early childhood) to cover the needs of the child's growing body **(UNICEF, 2019).**

- **Malnutrition**

A general term often used instead of undernutrition or under-nutrition, although technically it also refers to over-nutrition. A person is malnourished if their diet does not contain the nutrients required for growth or good health, or if they cannot fully assimilate the food they eat because of illness (undernutrition). They are also malnourished if they consume too many calories (over-nutrition).

Types of malnutrition

- **Moderate acute malnutrition**

Moderate acute malnutrition occurs when a child weighs no more than 70-80% of the average weight for a child of his or her height. Their brachial perimeter is between 110 and 119 millimetres.

- **Severe acute malnutrition**

This is the most serious stage of malnutrition. Severe acute malnutrition occurs when a child weighs less than 70% of the average weight for a child of his or her height. They have a brachial circumference of less than 110 mm, look very thin and sometimes have oedema on their arms. skin.

I.2 EMPIRICAL REVIEW

It is between conception and the age of three that a child's organs and tissues, brain and bones are formed, and their physical and cognitive potential is shaped. Malnutrition weakens the immune system, making children vulnerable to infections (malaria, diarrhoea, etc.), increasing the severity of illnesses and slowing healing. These illnesses in turn lead to worsening malnutrition. When treatment is delayed too long, it leads to long-term development handicaps.

I.2.1 Little or no income.

The results of the EDSM IV 2006 show that the risk of a child dying is higher for children living in the poorest households (124 per thousand) than for those living in wealthier households (80 per thousand).

I.2.2 Very low or non-existent level of education.

Studies in Mali have shown a significant relationship between the high level of chronic malnutrition in children and the low level of education of their mothers. According to the EDSMV 2012-2013, the level of chronic malnutrition is 40% among children of mothers with no education, 31% among those whose mothers have primary education, and 18% among those whose mothers have secondary education or more.

I.2.3. Ideation characteristics

Ideation is a set of psycho-graphic variables that determine whether or not a given behaviour is adopted. These variables include the knowledge, attitudes and practices associated with that behaviour and access to information or membership of social networks/social context that do or do not favour that behaviour.

I.2.4. Knowledge of nutrition

- Not familiar with good nutrition practices.
- Uses traditional medicine.
- Does not know the benefits of colostrum.
- Does not know that fortified local foods are available.

I.2.5. Nutritional attitudes

- Trivialisation of the importance of food diversification.
- Nutrition does not take priority over other expenses.
- Give priority to your husband's diet.
- Good food is for the rich.

I.2.6. Nutritional practices

- Food bans.
- Nutrient-rich diet (soup for post-partum women).
- Supplementary food, but not up to standard (not introduced at the right time; not appropriate for the child; not varied).
- Breastfeeding, but not exclusively for six months (giving water, for example).
- Self-medication.
- Low attendance at health centres.
- Poor food hygiene

The period between birth and the age of 3 is characterised by rapid growth and organ maturation. It is also a period when children are most exposed to nutritional deficiencies, stunted growth and certain infections (gastrointestinal and respiratory infections, malaria, measles, etc.). Poor dietary practices at this age have dramatic short-term consequences for the health and survival of the child, but also long-term consequences for health, reproduction and intellectual capacity, compromising the educational and socio-professional future of these children. At this stage of their lives, children need a balanced diet to ensure optimal somatic and intellectual development and a good future at school and in the workplace.

I.2.7. Infant diet

I.2.8. From birth to 6 months:

The WHO and UNICEF Global Strategy for Infant and Young Child Feeding recommends exclusive breastfeeding for infants up to 6 months; in other words, breast milk and nothing but breast milk should be given at this age unless breastfeeding is contraindicated. In this case, artificial milk is preferred. Breast milk is perfectly suited to the child's needs at this stage. It provides daily calories, micronutrients and water, provided the mother is well nourished (infants receive nutrients from their mothers through breast milk. If the mother is not well nourished, the infant will not get the nutrients it needs). It has the particularity of evolving according to the infant's needs. So there are three stages in its development:

1. Colostrum: produced during the first 3 to 4 days after birth, it is characterised by its yellow colour, thick consistency and, above all, by the fact that it is very rich in immunoglobulins and vitamins. This phase is prolonged in the event of premature delivery, when it is richer in polyunsaturated fatty acids, which are essential for the cerebral development of the newborn.
2. Transitional milk: produced between 5ème and 14ème . It is thinner than colostrum, orange and richer in lactose, casein and fat. During the day, it gradually adapts.
3. Mature milk: produced from the 15th day onwards, it provides newborn babies with the nutrients they need to grow properly. It has a higher water content and its composition varies from one feed to the next. At the beginning of the feed, it is richer in water, mineral salts and lactose, with a low energy density: it has a thirst-quenching role at this time. Over the course of the feed, the energy density increases, ensuring that the baby's calorie requirements are met.

I.2.9. From 6 months

An infant's needs are no longer fully covered by breast milk. This is the age at which a safe and appropriate complementary food should be introduced. Diversification takes place in three stages:

I.2. 10. From 6 to 12 months: the diet should consist mainly of breast milk.

It is combined with :

• with water

• with vegetables

• with seasonal fruit

• infant flour, potatoes

• with meat, fish or eggs

N.B.: solid foods must be crushed before being eaten by the child, as they will have difficulty swallowing them.

I.2.11. From 12 to 24 months: breastfeeding is continued but at a slower rate. important.

Solid foods may be introduced in addition to those listed above. The diet may consist of :

• breast milk

• fruit

• vegetables

• cereals: rice, wheat, maize, etc.

• water

• meat, fish or eggs

1.5. From 24 to 36 months: the diet consists mainly of the family meal and snacks (fruit, yoghurt, bread spread, bean, millet or rice cakes, etc.) Breastfeeding may or may not be continued.

CHAPTER II

METHODOLOGY

II.1. FRAMEWORK OF THE STUDY

This study was carried out in the province of South Kivu, in the town of Bukavu, particularly in the Bagira Health Zone. This comprises 8 health areas. These areas will be visited to gather the necessary information from a survey questionnaire.

II.1.1 Geographical location,

It is bounded :

- ✓ North: via the Nyamuhinga river, Lake Kivu and the Kabare territory
- ✓ In the south: through the Ibanda commune and the Kabare territory
- ✓ East: Lake Kivu and the communes of Ibanda and Kadutu
- ✓ To the west by the Nyakakungulwe river and the Kabare territory

Health map of the Bagira health zone

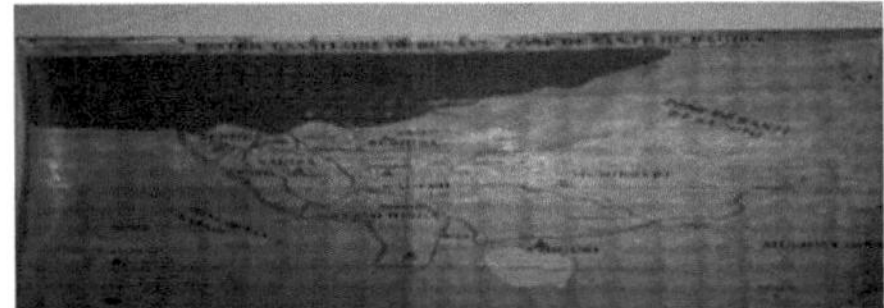

II.2 TYPE OF STUDY

This cross-sectional descriptive study aimed to describe mothers' knowledge, attitudes and practices with regard to complementary feeding in the Bagira Urban Health Zone. It took place between April and November 2023.

II.3 STUDY POPULATION

This study focused on breastfeeding women in the Bagira health zone. The target group was all mothers whose children were aged between 6 and 24 months.

A total of 384 women were surveyed.

II.4 CHOICE AND SIZE OF SAMPLE

II.4.1. Sample size

We used the LUNCH formula to find our sample size.

This formula enabled us to find the sample for our study.

$$\textit{Selon cette formule}, n = \frac{NZ\propto^2.p.(1-p)}{Nd^2 + Z\propto^2.p(1-p)}$$

$$= \frac{1\,57447.(1{,}96^2).0{,}5.(1-0{,}5)}{157447.(0{,}05) + (1{,}96^2).0{,}5.(1-0{,}5)} = 384$$

Hence :

n: sample size

P: Prevalence of the problem N : Total population

d: Corresponds to the margin of error, which is 5%.

Z: Coefficient corresponding to the 95% confidence level)

II.4.2. Sampling technique

This study will use proportional stratified random sampling, which involves finding the coefficient by dividing the sample size by the total population.

The coefficient is multiplied by the total number of employees in each AS in the ZS.

$$\text{Coef} = \frac{n}{N} = \frac{384}{157447} = 0\ ,0024$$

Table 1: Breakdown of the population by sample

Health areas/ Cs	N	n
BAGIRA	16677	40
BURHIBA	31581	76
CIGURHI	10560	29
KAHERO	16172	39
LUMU	20129	48
MAKOMA	13637	35
MUSHEKERE	26916	65
NYAMUHINGA	21775	52
Total	157447	384

II.5. ETHICAL CONSIDERATIONS

Prior to the survey and data collection, informed consent had to be sought and obtained. Participation in the study was free and unrestricted. The data was collected anonymously and the confidentiality of the results was ensured.

II.6. INCLUSION AND NON-INCLUSION CRITERIA

II.6.1. Inclusion criteria

All mothers living in the Bagira health zone who were present during the survey and who agreed to answer the questionnaire were included in the **study.**

II.6.2. Non-inclusion criteria

Any mother who does not meet the inclusion criteria mentioned above will be considered excluded from this study.

II.7. DATA COLLECTION TECHNIQUES AND TOOLS

II.7.1. Technical

Field visits to make contact with the target population in the Bagira ZS were the main technique used to collect data.

II.7.2 Data collection instrument

A well-developed survey questionnaire was used to collect the data found in the field. Data used in the problem and in the first chapter.

II.8. VARIABLES

II.8.1 Independent variables

- Age
- marital status
- level of education achieved
- religion
- line of business
- Number of children under 5
- Household size
- Child's age (6 to 24 months)
- Child's gender
- Knowledge
- Attitude
- Practice
- And so on.

II.8.2 Dependent variables

- Supplementary feeding is the dependent variable in this study.

II.9. DATA ANALYSIS

The following software and programmes were used:

- Kobo collect, which helped us collect and encode the data

-MICROSOFT OFFICE WORD 2016: for entering and using the literature relating to this study

- MICROSOFFT OFFICE EXCELL 2016: for arranging tables

- EPI INFO: for data encoding and analysis

II.10 DIFFICULTIES ENCOUNTERED

During the course of this work, a number of difficulties were encountered, such as :

- Mothers are sometimes absent from households, but have left their babies behind.
- Some mothers complained about the amount of time they spent with them.

CHAPTER III

PRESENTATION AND DISCUSSION OF RESULTS

III. 1. PRESENTATION OF THE RESULTS

Table 2: Socio-demographic characteristics of respondents.

Variables	n(384)	%	Average
Age>20	70	18.23	76.8
20-25 years	104	27.08	
25-30 years	98	25.52	
30-35 years old	60	15.62	
35-40	52	13.54	
Religion			
Catholic	255	66.41	
Protestant	77	20.05	
Muslim woman	23	5.99	
Jehovah's Witness	21	5.47	
Kimbanguist	8	2.08	
Level of study			
No level	31	8.07	
Primary	85	22.14	
Secondary	212	55.21	
University	56	14.58	
Profession			
No profession	240	62.5	
Retailer	83	21.61	
Employee	49	12.76	
Grower	12	3.12	

The table shows the large number of women surveyed, the majority of whom are aged between 20 and 25, the Catholic religion predominates (66.41%), 55.21% have a secondary education and 62.5% of the women surveyed have no occupation.

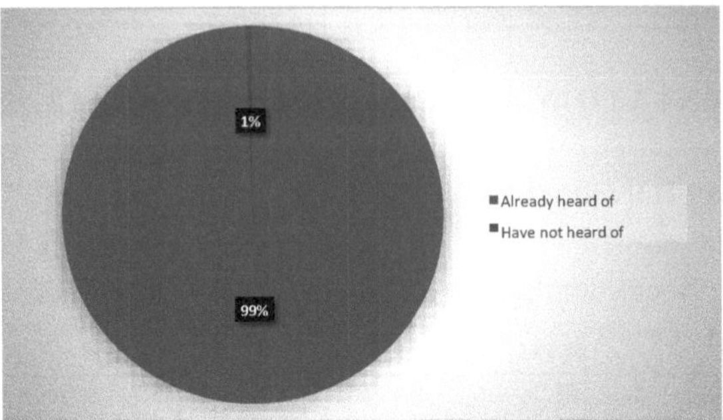

Figure 1: Distribution of respondents according to whether they have ever heard of complementary feeding

According to this figure, the majority of respondents (99%) had already heard of supplementary feeding.

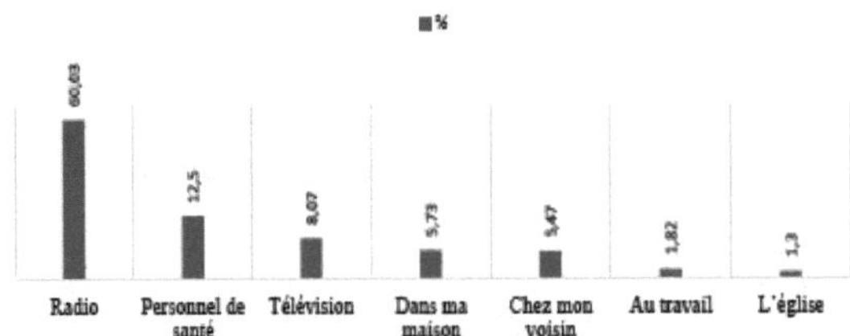

Figure 2: Breakdown of respondents by source or channel of information on complementary feeding.

According to this figure, most of the respondents (60.03%) get their information literally from the radio.

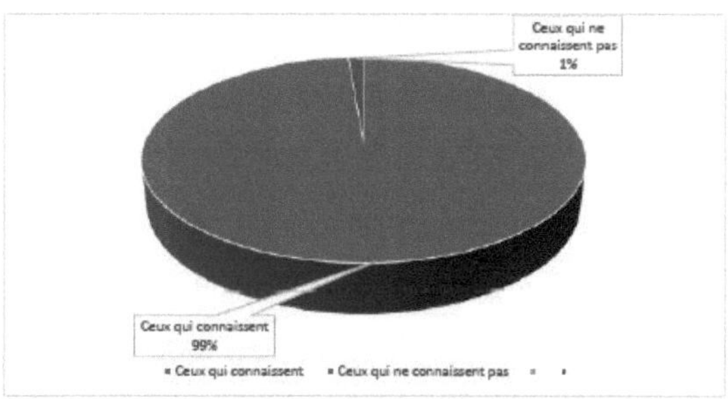

Figure 3: Distribution of respondents according to their knowledge of supplementary feeding

This figure shows that almost all of the respondents (99%) were familiar with complementary feeding as the gradual introduction of food to an infant.

Table 3: **Distribution of respondents according to their knowledge of supplementary feeding** v

Variables	n(384)	%
Knowledge of age of introduction of supplementary feed		
Have knowledge	383	99,6
Do not have the knowledge	1	0,4
If yes, age known		
From 6 months	383	100,0
Types of food to be introduced		
Ordinary porridge	362	91,8
Enriched porridge	21	7,8
Family dish	1	0,4
Weaning age		
I don't know	4	0,8
12 to 23 months	4	0,8
24 months or more	375	97,9
Other age	1	0,4

Reason for donating supplementary food

Breast milk alone is not enough 384 100,0

As well as breast milk, babies need other foods 384 100,0

Almost all of the respondents had already heard of complementary feeding and knew that it meant introducing food to supplement breast milk at 6 months. Ordinary porridge was identified as the very first food to be introduced by 9 out of 10 respondents, and almost all mentioned the age of over 24 months as the weaning age. The reasons for giving complementary food were that breast milk alone was not enough or that the baby needed other foods in addition to breast milk.

Table 4: Breakdown of respondents by mother's attitudes to complementary feeding

Variables	n(384)	%
Personal opinion on the AME		
Totally disagree	1	0,4
Disagreement	1	0,4
Neutral	35	9,4
Agreed.	347	86,9
Very much in agreement	7	2,9
Difficulties in preparing children's food		
Have difficulties	337	84,7
Do not have difficulties	47	15,3
Personal opinion on food diversification		
It's good for you	384	100,0
Difficulties in diversifying baby's diet		
Have difficulty diversifying	384	100,0
Personal opinion on feeding your child several times a day		
It's good for you	382	99,2
It's not beneficial	2	0,8
Difficulties in feeding the child several times a day		
Have difficulties	383	99,6
Do not have difficulties	1	0,4
Personal opinion on continuing to breastfeed after 6 months		

t's good for you	382	99,2
It's not beneficial	2	0,8
Difficulties in continuing to breastfeed after 6 months		
Have difficulties	19	5,8
Do not have difficulties	365	94,2

Looking at this table, we see that more than 8 out of 10 respondents agree with exclusive breastfeeding up to 6 months. Eight of the respondents (84.7%) had difficulties preparing food for their child, and almost all of them thought that diversifying their child's diet was beneficial, although they were aware of the difficulties involved in diversifying their child's diet. Almost all of them thought it was beneficial to feed their child several times a day, even though they found it difficult to do so. Almost all felt that continuing to breastfeed after 6 months was beneficial; 9 out of 10 had no problems doing so.

Table 5: Breakdown of respondents by complementary feeding practices

Variables	n = 384	%
Child's diet the day before the survey		
Oils and fats	381	98,8
Cereals	379	98,3
Vegetables and tubers rich in vitamin A	364	93,4
Sweets	364	93,4
Fish	352	88,1
Other vegetables	351	87,7
Other fruit	279	75,7
Dark green leafy vegetables	191	49,8
White tubers and roots	182	40,7
The Eggs	143	31,7
Meat	87	19,8
Milk and dairy products	46	14,4
Fruits Rich in Vitamin A	16	4,9
Iron-rich offal	8	2,9

The children's diet consisted mainly of oils and fats, cereals, vegetables and tubers rich in vitamin A, sweets, fish and fruit.

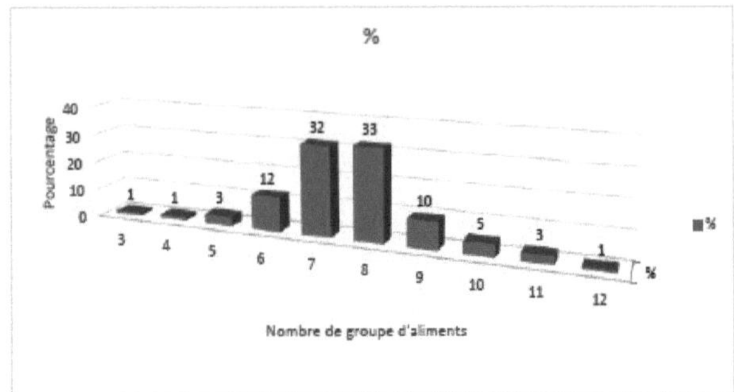

Figure 4: Distribution of respondents according to dietary diversity scores

Two-thirds of the women surveyed had given their children 7 to 8 food groups.

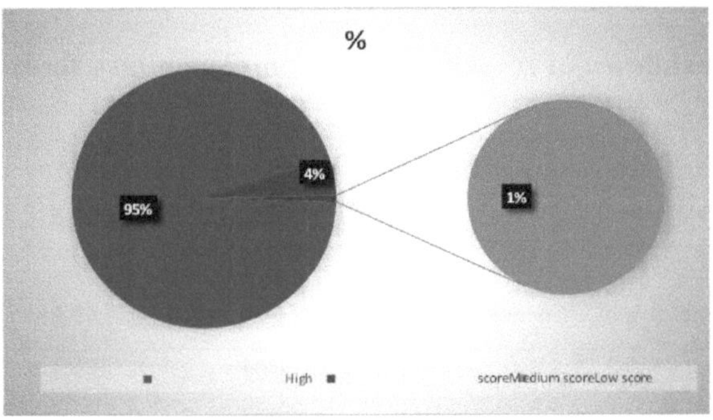

Figure 5: Classification of dietary diversity scores

Most of the respondents had a high dietary diversity score; more than 6 food groups.

III.2. DISCUSSION OF THE RESULTS.

Complementary foods must provide energy and nutrients of sufficient quality and quantity to supplement those provided by breast milk from the age of 6 months, thereby ensuring the child's optimal growth, development and good health **(Songré-Ouattara, L. et al. 2016)**.

Socio-demographic characteristics

This study involved 384 respondents and showed that 99% of them had already heard of complementary feeding. The main source of information was the radio (60.03%). Almost all the respondents knew that complementary feeding is the introduction of food to supplement breast milk at 6 months. Ordinary porridge was identified as the very first food to be introduced to the child by 9 out of 10 respondents, i.e. 91.8%, and almost all, i.e. 97.9%, mentioned the age of over 24 months as the weaning age. These results are almost identical to those of several studies. Table 2 shows that 7 out of 10 respondents were aged between 25 and 30. This is justified by the fact that the majority of breastfeeding women in the Democratic Republic of Congo are young. More than 9 out of 10 were married and almost a third had a secondary education. This would seem to be due to the fact that the level of literacy remains low despite the efforts made by the Congolese government in introducing the free primary education programme. One in 2 respondents was Catholic, and most were farmers. This is justified by the fact that the Congolese population is predominantly Christian and also by the fact that in rural areas, the majority of the population earns its living from farming.

Our results are similar to those of **Emmanuel B. Ngoy et al. Ngoy et al,** "Indicators of infant and young child feeding as predictors of malnutrition in children aged 6-23 months in the Kapolowe health zone, Haut-Katanga, DR Congo", a study showing that most of the children's mothers were married

(95.5%) **(Emmanuel B et all 2022).** Our results are similar to those of **Ms Korotimi SANOGO** in her study of mothers' knowledge and practices in terms of early childhood nutrition in the village of Point G in Commune III of the district of Bamako, a study which shows us that the age group between 25 and 34 was the most represented **(Ms Korotimi, 2011).**

These results differ from those of **Adama Moussa DIALLO**, who found that the 15-24 age group was in the majority, at 51.5%; the mothers were uneducated in 60.7% of cases, and 88.4% were housewives **(Diallo, A.M 2020).**

The majority of the children were aged between 6 and 12 months (55.1%), with an average age of 12.5 ± 5.2 months. This can be explained by the fact that this is the age group most affected by complementary feeding. Females predominated. This can be explained by the fact that in all populations, the proportion of female children is generally higher than the proportion of male children. More than 8 out of 10 households had two children under the age of 5. Efforts to raise awareness of birth spacing should be stepped up. And almost ¾ of households were made up of more than 7 people. This is due to the fact that, on average, households in the Democratic Republic of Congo are made up of 7 people.

1. Mothers' knowledge of supplementary feeding

In terms of the age at which other foods should be introduced, almost all the women knew the appropriate age at which complementary foods should be introduced (99.6%). All the women who said they knew enough about the age in question cited the 6th month as the ideal age for introducing other foods to the child. The table also shows that the type of food cited as the first to be introduced is ordinary porridge, cited by 91.8% of women. 97.94% of these women think that the child can already be weaned at 24 months or more. This table also describes the reasons for giving the child food other than breast milk.

According to all the women interviewed (100%), breast milk alone is not sufficient for the child's growth. According to all these women, in addition to breast milk, the baby needs other foods for its growth.

Our results in Table 3 differ from those of Adama Moussa DIALLO in his study of mothers' knowledge, attitudes and practices regarding the diet of children aged 0-23 months and their nutritional status in Niafunké, which shows that more than seven (7) out of ten (10), i.e. 72% of mothers, knew about AME up to 6 months, and the advantages of AME (73%). According to the same study, 85% of mothers were aware of the local foods used in complementary feeding. Knowledge of the quality and specificity of local foods is important for carrying out a affordable and achievable food diversification over time in a low-income population **(Diallo, A.M 2020).**

Our results corroborate those of **Korotimi SANOGO** in her study of mothers' knowledge and practices with regard to early childhood nutrition in the village of Point G in commune III of the Bamako district, who found that the mothers questioned had a fairly good knowledge of WHO recommendations with regard to appropriate early childhood nutrition practices. In fact, 98.9% thought that breast milk was the best food for newborns and infants; 68.1% knew the advantages of AM; 98.9% of AME; 86.6% knew that complementary food should be introduced from 6ème months, 98.2% knew what types of complementary food to give; 75.4% that breastfeeding should continue until 24 months or more **(Madame Korotimi, 2011)**.

To ensure optimal growth, health and development, the WHO recommends early initiation of breastfeeding for newborns, exclusive breastfeeding for up to 6 months, safe and appropriate dietary diversification from this age onwards, and continued breastfeeding for up to 24 months or beyond.

2. Mothers' attitudes to complementary feeding

Looking at table 4, we see that 86.9%, or the vast majority, think that it is more than necessary to breastfeed exclusively without introducing other foods, while 84.7% have difficulty preparing food for their children. They all think that it is beneficial to diversify the foods their children eat, but they all (100%) have difficulty making this diversification a reality.

The table also shows that 99.2% think it is beneficial to feed their child several times a day, although 99.6% find it difficult to do so. 99.2% think it is beneficial to continue breastfeeding their child after 6 months, while 5.8% find it difficult to continue breastfeeding their child after 6 months.

Data from the **Adama Moussa DIALLO** study showed that the vast majority of mothers were breastfeeding (94%). Our result is lower than this. This high rate of AM could be explained on the one hand by the fact that breastfeeding is a cultural practice and on the other hand by the fact that the majority of mothers do not have the means to obtain breast milk substitutes **(Diallo, A.M 2020).**

Our results converge with those of **Bougma S, Hama-Ba F, Garanet F et al**; Socio- demographic characteristics of mothers and complementary feeding practices among children aged 6 to 23 months in North Central Burkina Faso, which states that socio-demographic characteristics such as household size, child carer, number of children, area of residence, child age and mother's marital status are significantly associated with children's minimum dietary diversity in multivariate regression analyses. **(Bougma S et all, 2022).**

1. **Mothers' complementary feeding practices**

This section shows that 98.33% of the women had eaten cereals the day before the study was carried out, 93.42% had eaten vegetables and tubers rich in vitamin A, 40.66% had not eaten white tubers and roots, 49.79% had eaten dark green leafy vegetables and 87.65% had eaten other types of vegetables. Only 4.94% had given vitamin A-rich fruit and 75.72% had given other types of fruit.

In a study by **Bougma S, Hama-Ba F, Garanet F et al**; Socio-demographic characteristics of mothers and complementary feeding practices among children aged 6 to 23 months in North Central Burkina Faso, the positive impact of education on feeding practices was also noted in several studies. The mother's place of residence is another factor influencing the adequacy of minimum meal frequency practices. The study shows that children of urban mothers are 2.2 times more likely to have an adequate minimum meal frequency than their rural counterparts (OR=2.2; p=0.006). Mothers living in urban areas are more exposed to the various media that convey awareness-raising messages about infant and young child feeding practices. In addition, the majority of urban mothers are educated and understand the importance of these messages **(Bougma S et all, 2022).**

The table showing the types of food given to children shows that 2.88% had given their children offal rich in iron, 19.75% had given their children meat, 31.69% had given their children eggs, 88.07% had given their children fish, 14.40% had given their children milk or dairy products, 98.77% had given their children oils and fats and 93.42% had given their children sweets.

Emmanuel B. Ngoy et al. In their study entitled "Infant and young child feeding indicators as predictors of malnutrition in children aged 6-23 months in the Kapolowe health zone, Haut-Katanga, DR Congo", this study, which is similar to our own, showed that imported foods (cerélac, Dlite, etc.) are given first after exclusive breastfeeding, whether or not this is respected, by almost one mother in two (46.2%), followed by cereals (maize, rice, etc.).) are given first after exclusive breastfeeding, whether or not this is respected, by almost one mother in two (46.2%), followed by cereals (maize, rice, etc.) given by more than a third of mothers (36.7%). More than six out of ten mothers used a mixture of maize meal + oil + sugar as a supplementary feed, while 1.8% of mothers used a mixture of maize meal + fish. Cornflour + Soya was used by 2.7% of mothers, while almost all of them (95.2%) said that corn and soya were available locally produced foods **(Ms Korotimi, 2011).**

CONCLUSION

Here we are at the end of our end-of-cycle work, carried out with a view to obtaining our graduate diploma, on the subject of mothers' knowledge, attitudes and practices with regard to complementary feeding in the BAGIRA health zone. A food supplement, as the name suggests, is used to supplement a normal diet. Its aim is to help our bodies maintain or even improve their health. It is intended for people who wish to supplement their intake of certain nutrients as a result of a particular lifestyle, or it can be used to correct nutritional deficiencies or maintain an adequate intake of certain nutrients. What is the level of knowledge about malnutrition among the population of the BAGIRA health zone?

- What are the causes and consequences of malnutrition?

Beforehand, we set ourselves the overall objective of evaluating the knowledge, attitudes and practices of mothers in this Health Zone with regard to complementary food; having addressed a subject dealing with these points.

This was a cross-sectional descriptive study, and the LUNCH formula enabled us to find a sample of 384 breastfeeding mothers to survey in the area; and the data were collected using a survey questionnaire within the BAGIRA health zone. Our study showed that 99% of respondents had already heard of complementary feeding. The main source of information was the radio (60.03%). Almost all of the respondents knew that complementary feeding is the introduction of food to supplement breast milk at 6 months. Ordinary porridge was identified as the very first food to be introduced by 9 out of 10 respondents, i.e. 91.8%, and almost all (97.9%) mentioned the age of over 24 months as the weaning age. There is still work to be done in our country to improve feeding practices among young children in order to guarantee them better nutritional status, reduce morbidity and mortality rates linked to inappropriate feeding practices and thus ensure that these children have a more secure future at school and in the workplace. It is therefore important to improve women's knowledge of the composition of meals to be given to children so that they have a balanced diet.

RECOMMENDATIONS

At the end of our work, we are called upon to make recommendations that will be addressed both to healthcare staff and to mothers.

To healthcare staff :

- Advise mothers to practise early breastfeeding after giving birth by encouraging them to practise AME and introduce food at 6 months, while continuing to breastfeed until the child is 2 years old or more.
- Include the families of mothers (husbands, widows) and traditional birth attendants in awareness-raising activities on optimal child feeding practices and cooking demonstrations.

To mothers:

- To participate frequently in SPC sessions to improve feeding practices among young children in order to guarantee them better nutritional status, reduce morbidity and mortality rates linked to inappropriate feeding practices and thus ensure that these children have a more secure future at school and at work.
- Introduce adequate complementary food from the age of six months, while continuing to breastfeed until the age of two or more;

BIBLIOGRAPHICAL REFERENCES

2. Célestin KYAMBIKWA, health demography course, second year of public health degree, UOB, 2020-2021

3. Anonymous, Le Petit Larousse illustré, 1995, 1782p. Larousse Paris 1995. ISBN 2-03- 301195-X

4. Ministère de la Zone de santé, République démocratique du Congo, Recueil de Normes de la Zone de santé, August 2006

5. Essi M, njoyaoudou : l'enquête CAP (Connaissances, Attitudes, Pratiques) en recherche medical, Laboratoire de recherche sur les hepatites virales et communication en santé-FMSB correspondance, 2013, Dr Marie José Essi, laborhcs@gmail.com

6. Bougma S, Hama-Ba F, Garanet F et al ; Socio-demographic characteristics of mothers and complementary feeding practices among children aged 6 to 23 months in North Central Burkina Faso. African Journal of Food, Agriculture, Nutrition and Development. 2022

7. PRONANUT, Final report October 2022

8. Diallo, A.M. (2020) Connaissances, attitudes et pratiques des mères sur l'alimentation des enfants 0 à 23 mois et leur statut nutritionnel à la pediatrics/URENI of the CSRéf of Niafunké from December 2018 to February 2019. Thesis. Université des Sciences, des Techniques et des Technologies de Bamako. Available at: https://www.bibliosante.ml/handle/123456789/4524 (Accessed: 16 August 2023)

9. UNICEF. (2019). The State of Food Security and Nutrition in the World.

10. Salif Samaké et all, (2006) Cellule de Planification et de Statistique Ministère de la Santé Direction Nationale de la Statistique et de l'Informatique Ministère de l'Économie, de l'Industrie et du Commerce Bamako, Mali "Enquête

Démographique et de Santé du Mali (EDSM)

11. Emmanuel B. Ngoy, Ali M. Mapatano et al (2022) 'Infant and young child feeding indicators as predictors of malnutrition in children aged 6-23 months in the Kapolowe health zone, Haut-Katanga, DR Congo'.

12. UNHCR. (2020). World Food Security.

13. Songré-Ouattara, L. et al. (2016) 'Evaluation de l'aptitude nutritionnelle des aliments utilisés dans l'alimentation complémentaire du jeune enfant au Burkina Faso', Journal de la Société Ouest-Africaine de Chimie, 041, pp. 41-50.

14. Dr Lalla DIARRA 2020 - 2021Nutritional status of children aged 6 to 59 months and people aged 50 and over in IDP sites in Mali

15. ECHO Fact Sheet - Nutrition - May 2017

APPENDIX

SURVEY QUESTIONNAIRE

Our name is and we are third-year students studying for a degree in Public Health at the Official University of Bukavu. We have come to you to gather information in connection with our final year project on "**Mothers' knowledge, attitudes and practices regarding supplementary feeding in the health zone**". The information you give us will be strictly confidential. Thank you for your participation in our study.

I. **Socio-demographic characteristics**

1. How old are you?//

2. What is your marital status?

1. Married //2. Widowed //3. Divorced //4. Separated//

3. What is the highest level of education you have achieved?

1. none //2. Primary school //3. Secondary school // 4. University / / /

4. What religion are you?

1. No religion //2. Catholic //3. Protestant //4. Muslim // 5. Jehovah's Witness //

6. Other to be specified ....

5. What is your business sector?

1. Trade //2. Business services //3. Agriculture/

4. Household //5. No profession //6. Other to be specified ....

6. Residence ....

7. Number of children under 5//

8. Household size//

9. Child's age (6 to 24 months)//

II. mother's knowledge of complementary feeding

10. Gender of the child1. Male //2. Female/

11. Do you know the age at which other foods in addition to breast milk

should be introduced into your child's diet?1. Yes //**2. No //**

12. If yes, at what age? 1. before 6 months //2. From 6 months //

13. What type of food should I introduce at first?

1. Enriched porridge //2. Ordinary porridge //3. Family meal //

14. Until what age is it recommended that mothers continue to breastfeed their babies?

1. Don't know / /2. 6 months or less //3. 6-11 months //4. 12-23 months //5 . 24 months or more //6. Other to specify

15. Why is it important to give babies other foods in addition to breast milk from the age of 6 months?

1. Breast milk alone is not enough1. Yes //2. No //

2. Milk alone cannot provide all the nutrients a baby needs to grow Yes / /2. No//

3. Baby needs more food in addition to breast milk 1. yes //

2. No //

4. Other to be specified

THE MOTHER'S ATTITUDE TO COMPLEMENTARY FEEDING

16. What is your opinion on the statement made by health professionals "Breastfeed your children exclusively for 6 months before introducing any other food"? 1. Totally disagree / / 2 .Disagree / /3 .

Neutral / /4. Agree / /5.Very agree / /

17. Do you have any problems preparing food for your child?

1. Yes / /2. No / /

18. Do you think that giving different types of food is beneficial for the child?

1. Yes //2. No / /

19. Do you find it difficult to give your child different types of food?

1. Yes //2. No //

20. Do you think that feeding your child several times a day is beneficial? 1. yes//**2. No / /**

21. Do you find it difficult to feed your child several times a day?

1. Yes //2. No / /

22. Do you think it is beneficial to continue breastfeeding beyond 6 months?

1. yes//**2. No //**

23. Do you find it difficult to continue breastfeeding beyond 6 months? 1. yes //

2. No //

Mother's complementary feeding practices

24. Please indicate what you gave your child yesterday (meals and snacks), whether during the day or at night, at home or away from home. Start with the first food or drink eaten in the morning.

n°	Question		Answer YES=1 NO=0
a	Did the child eat anything (meal or snack) outside the home yesterday?		
b	Food group	Examples	
1	Cereals	Bread, noodles, biscuits, cookies or any other food made from millet, sorghum, maize, rice, wheat + insert local foods, e.g. ugali, porridge or pasta and other cereals available locally	
2	Vegetables And Tubers Rich In Vitamin A	Pumpkin, carrot, squash or sweet potato with orange flesh + other vitamin A-rich vegetables available locally (e.g. peppers)	
3	White Tubers And Roots	White potatoes, white yams, manioc or root-based foods	
4	Dark Green Leafy Vegetables	Dark green leafy vegetables, including wild species + locally available vitamin A-rich leaves such as manioc leaves, etc.	

5	Other vegetables	Other vegetables (e.g. tomato, onion, aubergine), including wild species	
6	Fruit Rich In Vitamin A	Ripe mangoes, cantaloupe, dried apricots, dried peaches + other vitamin A-rich fruits available locally	
7	Other fruit	Other fruit, including wild fruit	
8	Offal (Rich In Iron)	Liver, kidney, heart or other offal and blood-based foods	
9	Meat	Beef, pork, mutton, goat, rabbit, wild game, chicken, duck or other birds	
10	Eggs		
11	Fish	Fresh or dried fish or shellfish	
12	Pulses, Nuts And Seeds	Beans, peas, lentils, seeds or derived foods	
13	Milk and Dairy Products	Milk, cheese, yoghurt or other dairy products	
14	Oils and fats	Oil, fats or butter added to food, or used for the caisson	
15	Sweets	Sugar, honey, sweetened soft drinks or sweet foods such as chocolate, sweets, etc.	
16	Spices, Condiments, Drinks	Spices (black pepper, salt), condiments (soy sauce, hot sauce), coffee, tea, alcoholic beverages OR local examples	

Thank you for taking part in this study!

Printed by Books on Demand GmbH, Norderstedt / Germany